# Nutt...
# Metr...
# in th...

COMPILED BY
HELEN PERRIN, B.Sc., M.Sc.


**MAIN SECTIONS**

WEIGHT · LENGTH · AREA ·
VOLUME & CAPACITY · ENERGY & POWER ·
TEMPERATURE · PERSONAL MEASUREMENTS ·
CLOTHING SIZES · SHOE SIZES ·
HOME DRESSMAKING · METRIC COOKERY ·
PAPER & BOOK SIZES · BED SIZES ·
WOOD & BOARD · TINS OF PAINT ·
CONVERSION TABLES

**FREDERICK WARNE & CO LTD**
London · New York

# CONTENTS

Whilst the Publishers have endeavoured to ensure the accuracy of this information, they cannot hold themselves responsible for any inconvenience or loss occasioned by any error therein.

© Frederick Warne & Co Ltd, London, England, 1974
ISBN 0 7232 1804 8

*Printed in Great Britain by Henry Stone & Son (Printers) Ltd.,
Banbury, Oxon.*
341.1174

# INTRODUCTION

The **Imperial System** of measurement, based on the yard for length and the pound for weight, has been used in the United Kingdom for many centuries. The **Metric System** was founded by the French during the French Revolution. The **Système International d'Unités** (International System of Units) is the modern form of the metric system finally agreed on at an international conference in 1960. The international symbol for this system is **SI**. SI is the result of the rationalization of the whole structure of metric derived units. The United Kingdom decided to adopt SI as the primary system of weights and measures because almost 90% of the world's population live in countries that use SI, or are committed to use it.

SI is based on six base-units:

| Quantity | Unit | Symbol |
|---|---|---|
| length | metre | m |
| mass | kilogramme | kg |
| time | second | s |
| electric current | ampere | A |
| temperature | kelvin | K |
| luminous intensity | candela | cd |

All other units are derived from these six. For example velocity is expressed in metres/second. Some units are given special names, e.g. the joule (J), watt (W).

Multiples and sub-multiples are obtained by adding prefixes to the unit names. The most commonly used are:

| | | |
|---|---|---|
| **mega** | (M) | meaning multiplied by a million |
| **kilo** | (k) | meaning multiplied by a thousand |
| **deci** | (d) | meaning divided by ten |
| **centi** | (c) | meaning divided by a hundred |
| **milli** | (m) | meaning divided by a thousand |
| **micro** | ($\mu$) | meaning divided by a million |

The names of all units when written in full are in small letters, with the exception of the temperature unit **C**elsius or **C**entigrade, e.g. joule. In abbreviated form only those units formed from a proper name have a capital letter e.g. J. No full stop is used after an abbreviation and no plural s is used. The correct symbols for square and cubic measure are, respectively, $m^2$, $km^2$, etc., and $m^3$, $cm^3$, etc. Rather than use a comma to divide groups of three digits, the continental practice and scientific practice in the UK is to use a gap to separate groups of three digits. (e.g. 12 345·678 99 instead of 12,345·67899)

By international agreement the United Kingdom, Europe and most other countries use the spellings metre, litre and kilogramme. In North America the spellings used are meter, liter and kilogram.

# WEIGHT

The basic unit for weight is the **kilogramme**. One gramme is a very small weight, about the weight of a 'Smartie'. In cooking and food packaging 25 grammes will be the basic unit.

## Approximate conversions
1 kilogramme = $2\frac{1}{5}$ lb
25 grammes is slightly less than 1 oz
50 kilogrammes = 1 cwt
1 tonne (metric ton) is slightly less than 1 ton

## Exact conversions
### Imperial to SI
| | | |
|---|---|---|
| 1 ounce | = | 28·350 g |
| 1 pound | = | 454 g |
| 1 stone | = | 6·350 kg |
| 1 quarter | = | 12·701 kg |
| 1 hundredweight | = | 50·802 kg |
| 1 ton | = | 1·016 t |

### SI to Imperial
| | | |
|---|---|---|
| 1 gramme (g) | = | 0·035 27 oz |
| 1 kilogramme (kg) | = | 2·205 lb |
| 1 tonne (t) ( =1000 kg) | = | 0·984 ton |

# LENGTH

The base unit for length is the **metre**. For small lengths in precision work the millimetre will be used, but for household and everyday measurements the centimetre will be used. The kilometre will be used to measure distances.

## Approximate Conversions

1 metre = 39 in
10 centimetres = 4 in
8 kilometres = 5 miles

## Exact conversions
### Imperial to SI

| | | |
|---|---|---|
| 1 inch | = | 25·4 mm |
| 1 foot | = | 0·3048 m |
| 1 yard | = | 0·9144 m |
| 1 chain | = | 20·117 m |
| 1 furlong | = | 0·2012 km |
| 1 mile | = | 1·6093 km |

### SI to Imperial

1 millimetre (mm) = 0·0394 in
1 centimetre (cm) = 0·3937 in

$$1 \text{ metre (m)} = \begin{cases} 39·37 \text{ in} \\ 3·2808 \text{ ft} \\ 1·094 \text{ yd} \end{cases}$$

$$1 \text{ kilometre (km)} = \begin{cases} 0·6214 \text{ mile} \\ 1093·6 \text{ yd} \end{cases}$$

# AREA

The derived unit for area is the **square metre**. For larger areas the **hectare** (ha) is used. (1 hectare = 10 000 m²)

## Approximate Conversions

10 square centimetres = $1\frac{1}{2}$ square inches

1 square metre = $1\frac{1}{5}$ square yards,

                    or slightly more than 10 square feet

1 square kilometre = $\begin{cases} 250 \text{ acres} \\ \frac{2}{5} \text{ square mile} \end{cases}$

1 hectare = $2\frac{1}{2}$ acres (about the size of a full-sized football pitch)

## Exact Conversions

### Imperial to SI

1 square inch  =  645·16 mm²

1 square foot  = $\begin{cases} 0·092 \ 9 \text{ m}^2 \\ 929·03 \text{ cm}^2 \end{cases}$

1 square yard  =  0·836 1 m²

1 acre         = $\begin{cases} 4046·86 \text{ m}^2 \\ 0·404 \ 7 \text{ ha} \end{cases}$

1 square mile  = $\begin{cases} 2·590 \text{ km}^2 \\ 258·999 \text{ ha} \end{cases}$

### SI to Imperial

1 square centimetre =  0·155 0 in²

1 square metre    = $\begin{cases} 1·196 \text{ yd}^2 \\ 10·764 \text{ ft}^2 \end{cases}$

1 hectare         =  2·471 acres

1 square kilometre = $\begin{cases} 247·1 \text{ acres} \\ 0·386 \ 1 \text{ sq. mile} \end{cases}$

# VOLUME AND CAPACITY

The derived unit for volume is the **cubic metre,** but volumes of liquids are usually expressed in **litres** and **millilitres.**

| | | |
|---|---|---|
| 1 litre (l) | = | 1 dm$^3$ |
| 1 millilitre (ml) | = | 1 cm$^3$ |

(A cubic centimetre is very commonly, but incorrectly, referred to as a cc. 1 cc = 1 ml)

## Approximate Conversions

1 litre = $\begin{cases} \frac{1}{5} \text{ gal} \\ 1\frac{3}{4} \text{ pints} \\ 35 \text{ fluid ounces} \end{cases}$

15 millilitres = $\begin{cases} 1 \text{ tablespoon} \\ \frac{1}{2} \text{ fluid ounce} \end{cases}$

30 millilitres = 1 fluid ounce

## Exact Conversions
### Imperial to SI

| | | |
|---|---|---|
| 1 cubic inch | = | 16·387 cm$^3$ |
| 1 cubic foot | = | 0·028 3 m$^3$ |
| 1 cubic yard | = | 0·764 6 m$^3$ |
| 1 fluid ounce | = | 28·4 ml |
| 1 gill | = | 142 ml |
| 1 pint | = | 0·568 l |
| 1 quart | = | 1·137 l |
| 1 gallon | = | 4·546 l |

**SI to Imperial**

1 cubic centimetre = 0·061 in³

1 cubic metre = $\begin{cases} 35·315 \text{ ft}^3 \\ 1·308 \text{ yd}^3 \end{cases}$

1 millilitre = $\begin{cases} 0·061 \text{ in}^3 \\ 0·007\ 04 \text{ gill} \\ 0·035 \text{ fl. oz} \end{cases}$

1 litre = $\begin{cases} 0·220 \text{ gal} \\ 1·760 \text{ pints} \\ 61·026 \text{ in}^3 \end{cases}$

1 hectolitre = 22·00 gal

**US Units**

1 US pint = $\begin{cases} \frac{4}{5} \text{ Imperial pint} \\ 16 \text{ fl. oz} \end{cases}$ = 0·473 l

1 US gallon = $\begin{cases} \frac{4}{5} \text{ Imperial gallon} \\ 8 \text{ US pints} \end{cases}$ = 3·785 l

# ENERGY AND POWER

The **joule** (J) is the SI unit for energy and work in all its forms including heat, electricity mechanical work and chemical energy stored in fuels and foods.

The **watt** (W) is the SI unit of power. 1 hp = 745·7 W

The electricity units volt and ohm are derived from the base unit the ampere. Electricity is measured by the kilowatt hour. 1 kilowatt hour = 3·6 megajoules

The calorie is a metric unit but not an SI unit. The **C**alorie which is used for describing the energy value of foods is equal to 1000 calories. 1 calorie = 4·186 joules, 1 Calorie = 4·186 kilojoules

# TEMPERATURE

The unit for everyday use is the degree **Celsius** or the degree **Centigrade** (°C). (In France Centigrade is a unit of angular measure so, to prevent confusion, the name Celsius is preferred.)

To convert degrees Fahrenheit (°F) to °C, deduct 32 and multiply by $\frac{5}{9}$.

To convert °C to °F, multiply by $\frac{9}{5}$ and add 32.

Examples of temperature on the Celsius scale:

| | |
|---|---|
| freezing-point of water | 0°C |
| boiling-point of water | 100°C |
| average body temperature | 37°C |
| cool weather | 10°C |
| mild weather | 15°C |
| warm weather | 20°C |
| heat wave | 25°C |

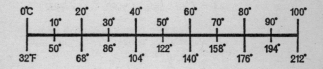

# PERSONAL MEASUREMENTS

## Weight

A person's weight will be quoted in kilogrammes. In practice it will probably be sufficient to quote one's weight as a whole number of kilogrammes, or to the nearest half kilogramme. (One pound is slightly less than half a kilogramme.) The table on pages 12–15 converts stones and pounds to kilogrammes, to one decimal place.

## Height

A person's height will be measured in metres, to two decimal places or, particularly for small children, height could be measured in centimetres. For example 5 ft 6 in converts to 1·68 m or 168 cm, and 3 ft to 91 cm. The table on pages 16–17 converts feet and inches to centimetres, for heights less than 4 ft, and to metres, for heights from 4 ft to 6 ft 5½ in.

## Body Measurements

Body measurements will be taken in centimetres. The following table converts inches to centimetres over the range of more common body measurements.

| in | cm | in | cm | in | cm |
|----|----|----|----|----|----|
| 20 | 51 | 30 | 76 | 40 | 102 |
| 21 | 53 | 31 | 79 | 41 | 104 |
| 22 | 56 | 32 | 81 | 42 | 107 |
| 23 | 58 | 33 | 84 | 43 | 109 |
| 24 | 61 | 34 | 86 | 44 | 112 |
| 25 | 64 | 35 | 89 | 45 | 114 |
| 26 | 66 | 36 | 91 | 46 | 117 |
| 27 | 69 | 37 | 94 | 47 | 119 |
| 28 | 71 | 38 | 97 | 48 | 122 |
| 29 | 74 | 39 | 99 | | |

| WEIGHT | | | stones, pounds to kilogrammes | | |
|---|---|---|---|---|---|
| stones lb | | kg | stones lb | | kg |
| 0 | 1 | 0·5 | 2 | 0 | **12·7** |
| | 2 | 0·9 | | 1 | 13·2 |
| | 3 | 1·4 | | 2 | 13·6 |
| | 4 | 1·8 | | 3 | 14·1 |
| | 5 | 2·3 | | 4 | 14·5 |
| | 6 | 2·7 | | 5 | 15·0 |
| | | | | 6 | 15·4 |
| 0 | 7 | **3·2** | 2 | 7 | **15·9** |
| | 8 | 3·6 | | 8 | 16·3 |
| | 9 | 4·1 | | 9 | 16·8 |
| | 10 | 4·5 | | 10 | 17·2 |
| | 11 | 5·0 | | 11 | 17·7 |
| | 12 | 5·4 | | 12 | 18·1 |
| | 13 | 5·9 | | 13 | 18·6 |
| 1 | 0 | **6·4** | 3 | 0 | **19·1** |
| | 1 | 6·8 | | 1 | 19·5 |
| | 2 | 7·3 | | 2 | 20·0 |
| | 3 | 7·7 | | 3 | 20·4 |
| | 4 | 8·2 | | 4 | 20·9 |
| | 5 | 8·6 | | 5 | 21·3 |
| | 6 | 9·1 | | 6 | 21·8 |
| 1 | 7 | **9·5** | 3 | 7 | **22·2** |
| | 8 | 10·0 | | 8 | 22·7 |
| | 9 | 10·4 | | 9 | 23·1 |
| | 10 | 10·9 | | 10 | 23·6 |
| | 11 | 11·3 | | 11 | 24·0 |
| | 12 | 11·8 | | 12 | 24·5 |
| | 13 | 12·2 | | 13 | 24·9 |

| stones, pounds to kilogrammes | | | WEIGHT | |
| --- | --- | --- | --- | --- |
| stones | lb | kg | stones lb | kg |
| 4 | 0 | 25·4 | 6 0 | 38·1 |
|   | 1 | 25·9 | 1 | 38·6 |
|   | 2 | 26·3 | 2 | 39·0 |
|   | 3 | 26·8 | 3 | 39·5 |
|   | 4 | 27·2 | 4 | 39·9 |
|   | 5 | 27·7 | 5 | 40·4 |
|   | 6 | 28·1 | 6 | 40·8 |
| 4 | 7 | 28·6 | 6 7 | 41·3 |
|   | 8 | 29·0 | 8 | 41·7 |
|   | 9 | 29·5 | 9 | 42·2 |
|   | 10 | 29·9 | 10 | 42·6 |
|   | 11 | 30·4 | 11 | 43·1 |
|   | 12 | 30·8 | 12 | 43·5 |
|   | 13 | 31·3 | 13 | 44·0 |
| 5 | 0 | 31·8 | 7 0 | 44·5 |
|   | 1 | 32·2 | 1 | 44·9 |
|   | 2 | 32·7 | 2 | 45·4 |
|   | 3 | 33·1 | 3 | 45·8 |
|   | 4 | 33·6 | 4 | 46·3 |
|   | 5 | 34·0 | 5 | 46·7 |
|   | 6 | 34·5 | 6 | 47·2 |
| 5 | 7 | 34·9 | 7 7 | 47·6 |
|   | 8 | 35·4 | 8 | 48·1 |
|   | 9 | 35·8 | 9 | 48·5 |
|   | 10 | 36·3 | 10 | 49·0 |
|   | 11 | 36·7 | 11 | 49·4 |
|   | 12 | 37·2 | 12 | 49·9 |
|   | 13 | 37·6 | 13 | 50·3 |

| WEIGHT | | | stones, pounds to kilogrammes | | |
|---|---|---|---|---|---|
| stones | lb | kg | stones | lb | kg |
| 8 | 0 | 50·8 | 10 | 0 | 63·5 |
|  | 1 | 51·3 |  | 1 | 64·0 |
|  | 2 | 51·7 |  | 2 | 64·4 |
|  | 3 | 52·2 |  | 3 | 64·9 |
|  | 4 | 52·6 |  | 4 | 65·3 |
|  | 5 | 53·1 |  | 5 | 65·8 |
|  | 6 | 53·5 |  | 6 | 66·2 |
| 8 | 7 | 54·0 | 10 | 7 | 66·7 |
|  | 8 | 54·4 |  | 8 | 67·1 |
|  | 9 | 54·9 |  | 9 | 67·6 |
|  | 10 | 55·3 |  | 10 | 68·0 |
|  | 11 | 55·8 |  | 11 | 68·5 |
|  | 12 | 56·2 |  | 12 | 68·9 |
|  | 13 | 56·7 |  | 13 | 69·4 |
| 9 | 0 | 57·2 | 11 | 0 | 69·9 |
|  | 1 | 57·6 |  | 1 | 70·3 |
|  | 2 | 58·1 |  | 2 | 70·8 |
|  | 3 | 58·5 |  | 3 | 71·2 |
|  | 4 | 59·0 |  | 4 | 71·7 |
|  | 5 | 59·4 |  | 5 | 72·1 |
|  | 6 | 59·9 |  | 6 | 72·6 |
| 9 | 7 | 60·3 | 11 | 7 | 73·0 |
|  | 8 | 60·8 |  | 8 | 73·5 |
|  | 9 | 61·2 |  | 9 | 73·9 |
|  | 10 | 61·7 |  | 10 | 74·4 |
|  | 11 | 62·1 |  | 11 | 74·8 |
|  | 12 | 62·6 |  | 12 | 75·3 |
|  | 13 | 63·0 |  | 13 | 75·7 |

| stones, pounds to kilogrammes | | | WEIGHT | | |
|---|---|---|---|---|---|
| stones | lb | kg | stones | lb | kg |
| 12 | 0 | **76·2** | 14 | 0 | **88·9** |
|  | 1 | 76·7 |  | 1 | 89·4 |
|  | 2 | 77·1 |  | 2 | 89·8 |
|  | 3 | 77·6 |  | 3 | 90·3 |
|  | 4 | 78·0 |  | 4 | 90·7 |
|  | 5 | 78·5 |  | 5 | 91·2 |
|  | 6 | 78·9 |  | 6 | 91·6 |
| 12 | 7 | **79·4** | 14 | 7 | **92·1** |
|  | 8 | 79·8 |  | 8 | 92·5 |
|  | 9 | 80·3 |  | 9 | 93·0 |
|  | 10 | 80·7 |  | 10 | 93·4 |
|  | 11 | 81·2 |  | 11 | 93·9 |
|  | 12 | 81·6 |  | 12 | 94·3 |
|  | 13 | 82·1 |  | 13 | 94·8 |
| 13 | 0 | **82.6** | 15 | 0 | **95·3** |
|  | 1 | 83·0 |  |  |  |
|  | 2 | 83·5 | 15 | 7 | **98·4** |
|  | 3 | 83·9 |  |  |  |
|  | 4 | 84·4 | 16 | 0 | **101·6** |
|  | 5 | 84·8 |  |  |  |
|  | 6 | 85·3 | 16 | 7 | **104·8** |
| 13 | 7 | **85·7** | 17 | 0 | **108·0** |
|  | 8 | 86·2 |  |  |  |
|  | 9 | 86·6 | 17 | 7 | **111·1** |
|  | 10 | 87·1 |  |  |  |
|  | 11 | 87·5 | 18 | 0 | **114·3** |
|  | 12 | 88·0 |  |  |  |
|  | 13 | 88·5 | 18 | 7 | **117·5** |

| HEIGHT | | | feet, inches to centimetres | | |
|---|---|---|---|---|---|
| ft | in | cm | ft | in | cm |
| 1 | 6 | 46 | 2 | 9 | 84 |
| | 6½ | 47 | | 9½ | 85 |
| | 7 | 48 | | 10 | 86 |
| | 7½ | 50 | | 10½ | 88 |
| | 8 | 51 | | 11 | 89 |
| | 8½ | 52 | | 11½ | 90 |
| 1 | 9 | 53 | 3 | 0 | 91 |
| | 9½ | 55 | | ½ | 93 |
| | 10 | 56 | | 1 | 94 |
| | 10½ | 57 | | 1½ | 95 |
| | 11 | 58 | | 2 | 97 |
| | 11½ | 60 | | 2½ | 98 |
| 2 | 0 | 61 | 3 | 3 | 99 |
| | ½ | 62 | | 3½ | 100 |
| | 1 | 64 | | 4 | 102 |
| | 1½ | 65 | | 4½ | 103 |
| | 2 | 66 | | 5 | 104 |
| | 2½ | 67 | | 5½ | 105 |
| 2 | 3 | 69 | 3 | 6 | 107 |
| | 3½ | 70 | | 6½ | 108 |
| | 4 | 71 | | 7 | 109 |
| | 4½ | 72 | | 7½ | 110 |
| | 5 | 74 | | 8 | 112 |
| | 5½ | 75 | | 8½ | 113 |
| 2 | 6 | 76 | 3 | 9 | 114 |
| | 6½ | 77 | | 9½ | 116 |
| | 7 | 79 | | 10 | 117 |
| | 7½ | 80 | | 10½ | 118 |
| | 8 | 81 | | 11 | 119 |
| | 8½ | 83 | | 11½ | 121 |

| feet, inches to metres | | | HEIGHT | | |
|---|---|---|---|---|---|
| ft | in | m | ft | in | m |
| **4** | **0** | **1·22** | **5** | **3** | **1·60** |
| | ½ | 1·23 | | 3½ | 1·61 |
| | 1 | 1·24 | | 4 | 1·63 |
| | 1½ | 1·26 | | 4½ | 1·64 |
| | 2 | 1·27 | | 5 | 1·65 |
| | 2½ | 1·28 | | 5½ | 1·66 |
| **4** | **3** | **1·30** | **5** | **6** | **1·68** |
| | 3½ | 1·31 | | 6½ | 1·69 |
| | 4 | 1·32 | | 7 | 1·70 |
| | 4½ | 1·33 | | 7½ | 1·71 |
| | 5 | 1·35 | | 8 | 1·73 |
| | 5½ | 1·36 | | 8½ | 1·74 |
| **4** | **6** | **1·37** | **5** | **9** | **1·75** |
| | 6½ | 1·38 | | 9½ | 1·77 |
| | 7 | 1·40 | | 10 | 1·78 |
| | 7½ | 1·41 | | 10½ | 1·79 |
| | 8 | 1·42 | | 11 | 1·80 |
| | 8½ | 1·44 | | 11½ | 1·82 |
| **4** | **9** | **1·45** | **6** | **0** | **1·83** |
| | 9½ | 1·46 | | ½ | 1·84 |
| | 10 | 1·47 | | 1 | 1·85 |
| | 10½ | 1·49 | | 1½ | 1·87 |
| | 11 | 1·50 | | 2 | 1·88 |
| | 11½ | 1·51 | | 2½ | 1·89 |
| **5** | **0** | **1·52** | **6** | **3** | **1·91** |
| | ½ | 1·54 | | 3½ | 1·92 |
| | 1 | 1·55 | | 4 | 1·93 |
| | 1½ | 1·56 | | 4½ | 1·94 |
| | 2 | 1·57 | | 5 | 1·96 |
| | 2½ | 1·59 | | 5½ | 1·97 |

# CLOTHING SIZES

The commonest size interval used in the UK is 2 inches, which is slightly more than 5 centimetres. Unfortunately if the common range of sizes for bust or chest measurements is converted to the nearest centimetre equivalents, a 5 cm interval does not occur throughout the range.

| in | nearest cm equivalent |
|:---:|:---:|
| 32 | 81 |
| 34 | 86 |
| 36 | 91 |
| 38 | 97 |
| 40 | 102 |
| 42 | 107 |
| 44 | 112 |
| 46 | 117 |
| 48 | 122 |

At present this table has been widely adopted as the standard for converting Imperial to metric sizes. It is envisaged that, at a much later date, a system with equal intervals will be adopted. For women's wear the interval will be 5 cm and for men's wear 4 cm.

**Men's Wear**
Initially the Imperial size will be converted to the nearest whole centimetre, without any change in the actual dimensions of the garment. Both sizes will be given on the label with the metric conversion first, e.g. 97 cm/38 inch chest. This means that where a 2 inch size interval is used the metric conversion shown above will be used. Eventually it is recommended that manufacturers of men's wear should follow continental practice and adopt a 4 cm size interval, including 100 cm. The label will then show the imperial conversion to the nearest half-inch, e.g. 108cm/42½ inch. The following table shows the size range based on this recommendation.

| 4 cm intervals | Imperial equivalents to nearest $\frac{1}{2}$ in |
|---|---|
| 80 | $31\frac{1}{2}$ |
| 84 | 33 |
| 88 | $34\frac{1}{2}$ |
| 92 | 36 |
| 96 | 38 |
| 100 | $39\frac{1}{2}$ |
| 104 | 41 |
| 108 | $42\frac{1}{2}$ |
| 112 | 44 |
| 116 | $45\frac{1}{2}$ |
| 120 | 47 |
| 124 | 49 |

Initially there will also be no change in the actual dimensions of shirts. Size markings will have the metric measurements first, followed by the present inch size, e.g. 38 cm–15. There is no inch size equivalent to sizes 35 cm, 40 cm and 45 cm, and it is recommended that these sizes should be marked in the way shown in the following table, e.g. 39–40 cm–15½. At a later stage it is envisaged that the sizes will be adjusted slightly and 35 cm, 40 cm and 45 cm will be included as separate sizes.

| Old size | Metric size designation |
|---|---|
| 13 | 33 cm |
| $13\frac{1}{2}$ | 34–35 cm |
| 14 | 36 cm |
| $14\frac{1}{2}$ | 37 cm |
| 15 | 38 cm |
| $15\frac{1}{2}$ | 39–40 cm |
| 16 | 41 cm |
| $16\frac{1}{2}$ | 42 cm |
| 17 | 43 cm |
| $17\frac{1}{2}$ | 44–45 cm |
| 18 | 46 cm |

## Women's Wear

The existing size codes will be retained. Garments designed specifically for shorter or taller women will be provided with a suffix to the size code of

S for heights up to 160 cm (5 ft 3 in),
T for heights of 170 cm (5 ft 7 in) or over.

The body measurements in centimetres which have been proposed for each size correspond very closely to the direct conversions of the body measurements in inches for that size, and the actual dimensions of a garment will be virtually unaltered. Garment labels will show the size code followed by the appropriate dimension in centimetres and the inch equivalent. The table shows the centimetre measurements for each size, as proposed in a British Standards draft for public comment (Autumn 1973), and the measurements in inches which they replace.

| Size Code | Bust | | Hips | |
|---|---|---|---|---|
| | Inches | Proposed cm equivalent | Inches | Proposed cm equivalent |
| 8 | 32–33 | 78–82 | 34–35 | 83–87 |
| 10 | 33–34 | 82–86 | 35–36 | 87–91 |
| 12 | 34–35½ | 86–90 | 36–37½ | 91–95 |
| 14 | 35½–37 | 90–94 | 37½–39 | 95–99 |
| 16 | 37–39 | 95–99 | 39–41 | 100–104 |
| 18 | 39–41 | 100–104 | 41–43 | 105–109 |
| 20 | 41–43 | 105–109 | 43–45 | 110–114 |

(The author is grateful to the Clothing Institute for the information they supplied for this section.)

# SHOE SIZES

## Mondopoint System

A Mondopoint size marking consists of two numbers, e.g. 240/92. The first number is called the **size,** the second is called the **width**. The size of a shoe is the length in millimetres of the average foot which fits the shoe. The width of a shoe is the width in millimetres of the average foot which fits the shoe. E.g. The shoe marked 240/92 fits a foot of length 240 mm and width 92 mm. Size intervals will be either 5 mm or $7\frac{1}{2}$ mm. The size interval of 5 mm is recommended for ladies' shoes and good quality men's and children's shoes. The size interval of $7\frac{1}{2}$ mm will be used for moulded footwear, lace-up shoes, slippers, Wellington boots, etc. The sizes will be multiples of $7\frac{1}{2}$ mm, but the odd half-millimetre will be omitted from the size marking. The use of the width marking is recommended, even for footwear made in only one width. In multiple-fitting ranges, the width interval between fittings will be 3 or 4 mm.

A typical range of ladies' shoes would be :

> 215/82, 220/84, 225/86, 230/88, 235/90, 240/92, 245/94, 250/96, 255/98, 260/100, 265/102, 270/104
> (Equivalent to $2\frac{1}{2}$–9 on the English scale.)

A typical range of men's shoes would be :

> 240/90, 247/93, 255/96, 262/99, 270/102, 277/105, 285/108, 292/111
> (Equivalent to $5\frac{1}{2}$–$11\frac{1}{2}$ on the English scale.)

The tables on page 22 give the sizes in millimetres which correspond to the English sizes. Thus a child with English size 10 foot will take a Mondopoint size 170, or perhaps 175, and an adult with size 5 feet of average width will probably find the best fit with a Mondopoint size 235/90 or 240/92.

(The author is grateful to the Shoe and Allied Trades Research Association for the information they supplied for this section.)

## Suggested Average Conversion Chart
(English to Mondopoint)

| Children's Sizes | |
|---|---|
| 4 | 125 |
| 4½ | 130 |
| 5 | 132 |
| 5½ | 137 |
| 6 | 140 |
| 6½ | 145 |
| 7 | 147 |
| 7½ | 152 |
| 8 | 157 |
| 8½ | 160 |
| 9 | 165 |
| 9½ | 167 |
| 10 | 172 |
| 10½ | 175 |
| 11 | 180 |
| 11½ | 185 |
| 12 | 187 |
| 12½ | 192 |
| 13 | 197 |
| 13½ | 200 |

| Adult's Sizes | |
|---|---|
| 1 | 205 |
| 1½ | 207 |
| 2 | 212 |
| 2½ | 217 |
| 3 | 222 |
| 3½ | 225 |
| 4 | 230 |
| 4½ | 235 |
| 5 | 237 |
| 5½ | 242 |
| 6 | 245 |
| 6½ | 250 |
| 7 | 255 |
| 7½ | 257 |
| 8 | 262 |
| 8½ | 267 |
| 9 | 270 |
| 9½ | 275 |
| 10 | 280 |
| 10½ | 285 |
| 11 | 287 |
| 11½ | 292 |
| 12 | 297 |

# HOME DRESSMAKING

## Textiles

It has been recommended that textiles should be supplied in widths in steps of 50 cm, giving cloth widths of 50 cm, 100 cm, 150 cm, etc. Where these first preferred widths are unsuitable, intermediate widths should be spaced at 10 cm steps, for example 60 cm, 70 cm, 80 cm. At the time of writing textiles are only supplied in the preferred metric widths which are almost the same as previously used Imperial widths:

$$36 \text{ in becomes } 90 \text{ cm}$$
$$54 \text{ in becomes } 140 \text{ cm}$$
$$60 \text{ in becomes } 150 \text{ cm}$$

The textile industry has begun to supply fabrics in metric lengths. The minimum measurement used will be $\frac{1}{10}$ metre (0·1 m or 10 cm) (4 in), as $\frac{1}{8}$ yard ($4\frac{1}{2}$ in) has been previously.

## Paper Patterns

The major pattern companies have agreed on the metric body measurements within each Size. These fall within the range of measurements given in the Clothing Sizes section. Body measurements in metric are given on patterns now, as is a metric chart giving the amount of fabric needed. The actual dimensions of a made-up garment within a particular Size are almost the same when metric measurements are used as when Imperial measurements are used.

## Sewing Machine Needles

In the metric system used in Europe the number of the needle is equal to the shank diameter expressed in hundredths of a millimetre, e.g. 80 has a shank diameter of 0·8 mm. The following table shows the metric equivalent of the most widely used British system (the Singer system).

| British | Metric |
|---------|--------|
| 11 | 70 |
| 12 | 80 |
| 14 | 90 |
| 16 | 100 |

## Machine Stitches

| stitches/in | stitches/cm |
|-------------|-------------|
| 14 | 6 |
| 12 | 5 |
| 10 | 4 |
| 8 | 3 |

## Hand Knitting

Most firms now produce yarn in metric packs.

20 g is almost equal to $\frac{3}{4}$ oz

25 g is slightly less than 1 oz

40 g is slightly less than $1\frac{1}{2}$ oz

50 g is almost equal to $1\frac{3}{4}$ oz

New patterns quote measurements in centimetres and inches.

## Crochet Hooks

Until July 1969 there were two ranges of crochet hooks available in the UK, a wool and a cotton range. Since July 1969 the International Standard range (ISR) has been used. The table shows the ISR equivalent of the old sizes. The ISR size is the diameter of the hook in millimetres.

| ISR hooks | old Wool size | ISR hook | old Cotton size |
|-----------|---------------|----------|-----------------|
| 7·00 | 2 | 2·00 | $1\frac{1}{2}$ |
| 6·00 | 4 | 1·75 | $2\frac{1}{2}$ |
| 5·50 | 5 | 1·50 | $3\frac{1}{2}$ |
| 5·00 | 6 | 1·25 | $4\frac{1}{2}$ |
| 4·50 | 7 | 1·00 | $5\frac{1}{2}$ |
| 4·00 | 8 | 0·75 | $6\frac{1}{2}$ |
| 3·50 | 9 | 0·60 | 7 |
| 3·00 | 10 | | |
| 2·50 | 12 | | |
| 2·00 | 14 | | |

## Zip Fasteners

| in | cm | in | cm | in | cm |
|----|----|----|----|----|----|
| 4 | 10 | 10 | 25 | 22 | 55 |
| 5 | 12 | 12 | 30 | 24 | 60 |
| 6 | 15 | 14 | 35 | 26 | 65 |
| 7 | 18 | 16 | 40 | 28 | 70 |
| 8 | 20 | 18 | 45 | 30 | 75 |
| 9 | 22 | 20 | 50 | | |

# METRIC COOKERY

Many housewives will probably continue for some time to use their existing recipes, scales, measuring and cooking vessels regardless of the metric packs in which ingredients are supplied. Nevertheless, the next generation of house-wives will have been educated to use metric units, new equipment will be designed to hold and measure metric quantities, and new recipes will be devised using metric quantities. Agreement has been reached, after much discussion, on the conversion of recipes based on Imperial units and the metrication of equipment.

One ounce is equal to 28·350 grammes and obviously direct conversion would result in very awkward metric quantities. It has been decided to base recipes on a **25 gramme unit** as the equivalent of one ounce. This has three main advantages over the main alternative, the 30 gramme unit.

1  The 25 g unit·is easy for multiplication, e.g. 4 oz converts to 100 g, 6 oz to 150 g, etc.

2  When recipes using eggs are converted, the results are better.

3  The converted quantities adapt better to existing equipment such as cake tins and scales.

For measuring liquids the litre and millilitre will be used, using the conversion 1 pint to ½ litre or 500 millilitres. One useful fact is that 1 ml of water weighs 1 g and this applies to milk, beaten egg, stock, etc., so in practice it will be unnecessary to weigh liquids.

Recipes which depend on proportions of egg to flour, fats, sugar, etc. will be easy to convert. For example, where 1 egg to every 2 oz flour has been used, then 1 egg to every 50 g flour is the basic metric recipe. Similarly, for milk puddings where 2 oz grain was used for every pint of milk, the basic metric recipe will use 50 g grain to ½ litre milk.

## Measuring Equipment

Formerly, terms such as tea-dessert-table-spoons, cups, pints, etc. have been used and these will all be replaced by expressing the required quantity in millilitres. There will be **spoons** to measure 1·25, 2·5, 5, 10, 15 and 20 ml. (Approximately: a teaspoon holds 5 ml, a dessertspoon holds 10 ml and a tablespoon holds 20 ml.) So recipes will read:

> 5 ml spoon baking powder
> 10 ml spoon milk
> 2 × 20 ml spoons jam, etc.

**Cups** will be available in 50, 75, 100, 150 and 300 ml sizes, and will be referred to as measures, not cups, to avoid confusion with tableware. **Measuring jugs** will be marked to show 100, 125, 250, 500 and 750 ml, and possibly intermediate amounts.

**Kitchen scales** will differ from those used for trade, which are calibrated at intervals of 10 g or 20 g; domestic scales will show 25 g intervals. Existing scales which require a set of weights can easily be metricated by purchasing a set of metric weights.

## Handy Measures

| | | |
|---|---|---:|
| Flour | 1 rounded 20 ml spoon (tablespoon) holds | 20 g |
| | 2 level 20 ml spoons (tablespoons) hold | 25 g |
| | 75 ml measure holds | 50 g |
| | 150 ml measure holds | 100 g |
| Sugar | 1 level 20 ml spoon (tablespoon) holds | 20 g |
| | 100 ml measure holds | 100 g |
| Rice | 1 level 20 ml spoon (tablespoon) holds | 20 g |
| | 1 rounded 20 ml spoon (tablespoon) holds | 40 g |
| Liquids (water, milk, stock, beaten egg) | 1 20 ml spoon (tablespoon) holds | 20 g |
| | 100 ml measure holds | 100 g |
| Syrup | 1 20 ml spoon (tablespoon) holds | 30 g |
| | 3 10 ml spoons (dessertspoons) hold | 50 g |
| | 3 20 ml spoons (tablespoons) hold | 100 g |

Fats will be sold in blocks of 250 g [replacing $\frac{1}{2}$ lb (227 g)]. These will divide into 10 cubes of 25 g.

## Cooking Times

A converted recipe will make up to a slightly smaller quantity; thus cooking times will be slightly less using existing Fahrenheit oven temperatures. For example, where 30 minutes cooking time is stated, 25 minutes should be adequate.

## Oven Temperatures

Oven temperatures will be calibrated in Celsius (Centigrade). The following table shows the recommended Celsius temperatures to replace the existing Fahrenheit values, and the equivalent Gas Marks. (The Celsius temperature is approximately half the Fahrenheit temperature.)

|  | Electric | | Gas |
|---|---|---|---|
|  | **Old** | **New** |  |
| Cool | 200°F | 95°C | $\frac{1}{4}$ |
|  | 225°F | 110°C |  |
| Slow | 250°F | 130°C | $\frac{1}{2}$ |
|  | 275°F | 140°C | 1 |
|  | 300°F | 150°C | 2 |
| Moderate | 325°F | 160°C | 3 |
|  | 350°F | 180°C | 4 |
| Fairly Hot | 375°F | 190°C | 5 |
|  | 400°F | 200°C | 6 |
| Hot | 425°F | 220°C | 7 |
|  | 450°F | 230°C | 8 |
| Very Hot | 475°F | 240°C | 9 |
|  | 500°F | 260°C |  |

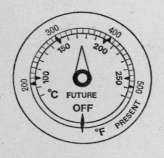

## Frying Temperatures

| | |
|---|---|
| 300°F | 150°C |
| 340°F | 170°C |
| 350°F | 175°C |
| 360°F | 180°C |
| 370°F | 190°C |
| 380°F | 195°C |
| 390°F | 200°C |

## Refrigerator and Freezer Temperatures

| | | | |
|---|---|---|---|
| | 7°C | 47°F | Average temperature in main cabinet of refrigerator |
| | 4°C | 40°F | |
| | 0°C | 32°F | Freezing-point of water |
| ☆ | −6°C | 21°F | One star frozen food compartment of refrigerator |
| ☆☆ | −12°C | 10°F | Two star frozen food compartment of refrigerator |
| ☆☆☆ | −18°C | 0°F | Three star frozen food compartment of refrigerator and food freezer at normal storage setting. |

# PAPER AND BOOK SIZES

**International (ISO) Sizes**
**'A' Series**
Stationery, Books and Magazines
(Trimmed or Finished sizes)

|      | mm               | in                      |
|------|------------------|-------------------------|
| A0   | 841 × 1189       | 33·11 × 46·81           |
| A1   | 594 × 841        | 23·39 × 33·11           |
| A2   | 420 × 594        | 16·54 × 23·39           |
| A3   | 297 × 420        | 11·69 × 16·54           |
| A4   | 210 × 297        | 8·27 × 11·69            |
| A5   | 148 × 210        | 5·83 × 8·27             |
| A6   | 105 × 148        | 4·13 × 5·83             |
| A7   | 74 × 105         | 2·91 × 4·13             |
| A8   | 52 × 74          | 2·05 × 2·91             |
| A9   | 37 × 52          | 1·46 × 2·05             |
| A10  | 26 × 37          | 1·02 × 1·46             |

In addition there is a **'B' Series,** with sizes intermediate between any two adjacent sizes of the 'A' series (for posters, etc.) and a **'C' Series** (for envelopes).

The sizes most commonly met with will be A4 and A5 which will gradually replace the traditional foolscap and quarto sizes.

# BED SIZES

There is a basic range of four metric bed sizes.

**Standard Single**
200 cm × 100 cm (6 ft 6¾ in × 3 ft 3⅜ in)

**Small Single**
190 cm × 90 cm (6 ft 2¾ in × 2 ft 11½ in)

**Standard Double**
200 cm × 150 cm (6 ft 6¾ in × 4 ft 11 in)

**Small Double**
190 cm × 135 cm (6 ft 2¾ in × 4 ft 5 in)

The Small Single and Small Double are almost the same as the former 3 ft × 6 ft 3 in and 4 ft 6 in × 6 ft 3 in single and double beds. The **Bunk Bed** size of 2 ft × 6 ft 3 in becomes 75 × 190 cm (approximately 2 ft 5½ in × 6 ft 2¾ in).

# WOOD AND BOARD

Softwood, and to an increasing extent hardwood, come from the mills in standard metric lengths and sections, and will be sold in metric measures. Where lengths were quoted in feet, they will now be given in metres. Where sections were quoted in inches, they will be given in millimetres.

### Lengths

**Sawn Softwood** The standard range begins at 1·8 metres and increases in steps of 300 millimetres to 6·3 metres. 300 mm is slightly shorter than 1 foot, and this range corresponds to 6 ft to 21 ft in steps of 1 foot.

**Sawn Hardwood** The standard range begins at 1·8 metres and increases in steps of 100 millimetres, corresponding to a range starting at 6 feet and increasing in steps of 4 inches.

## Standard Sections

Most of the metric sizes of sections are very close to the customary inch sizes. They are very slightly smaller.

**Sawn Softwood** The table shows the available sizes. The equivalent inch sizes are given in brackets.

| Thickness mm (equivalent inch size) | Width mm (equivalent inch size) | | | | | | |
|---|---|---|---|---|---|---|---|
| | 75 (3) | 100 (4) | 125 (5) | 150 (6) | 175 (7) | 200 (8) | 225 (9) |
| 16 ($\frac{5}{8}$) | x | x | x | x | | | |
| 19 ($\frac{3}{4}$) | x | x | x | x | | | |
| 22 ($\frac{7}{8}$) | x | x | x | x | | | |
| 25 (1) | x | x | x | x | x | x | x |
| 32 (1$\frac{1}{4}$) | x | x | x | x | x | x | x |
| 38 (1$\frac{1}{2}$) | x | x | x | x | x | x | x |
| 44 (1$\frac{3}{4}$) | x | x | x | x | x | x | x |
| 50 (2) | x | x | x | x | x | x | x |
| 63 (2$\frac{1}{2}$) | | x | x | x | x | x | x |
| 75 (3) | | x | x | x | x | x | x |
| 100 (4) | | x | | x | | x | |

**Sawn Hardwood** The following thicknesses (in mm) are available. The equivalent inch sizes are given in brackets.
19 ($\frac{3}{4}$), 25 (1), 32 (1$\frac{1}{4}$), 38 (1$\frac{1}{2}$), 50 (2), 63 (2$\frac{1}{2}$), 75 (3), 100 (4), 125 (5)

Widths are normally 150 mm and go up in steps of 10 mm. Widths of strips and narrows are 50 mm and go up in steps of 10 mm.

# TINS OF PAINT

Paint is sold in metric sized tins, marked in litres and millilitres. Tin diameters have not changed. Old brushes will fit new tins. The standard range of tin sizes agreed by the industry is:

**5 litres**    Replacing the 1 gallon tin.
       It contains 10% more.

**2·5 litres**   Replacing the ½ gallon tin.
       It contains 10% more.

**1 litre**     Replacing the quart tin.
       It contains 12% less.

**500 millilitres** Replacing the 1 pint tin.
       It contains 12% less.

**250 millilitres** Replacing the ½ pint tin.
       It contains 12% less.

**Coverage** The table shows the approximate spreading capacity in square metres on a non-porous surface. The figure in brackets shows the spreading capacity of the Imperial sized tin which that metric size replaces.

| Metric tin size | primer | gloss | emulsion |
|---|---|---|---|
| 5 litres | 60 (54) | 75 (67) | 90 (80) |
| 2·5 litres | 30 (27) | 37 (33) | 45 (40) |
| 1 litre | 12 (14) | 15 (17) | 18 (20) |
| 500 ml | 6 (7) | 7½ (8½) | 9 (10) |
| 250 ml | 3 (3½) | 3½ (4) | 4½ (5) |

| kilogrammes (kg) to pounds (lb) | | WEIGHT | |
| --- | --- | --- | --- |
| kg | lb | kg | lb |
| 0·1 | 0·2205 | 100 | 220·462 |
| 0·2 | 0·4409 | 150 | 330·693 |
| 0·3 | 0·6614 | 200 | 440·925 |
| 0·4 | 0·8818 | 250 | 551·156 |
| 0·5 | 1·1023 | 300 | 661·387 |
| 0·6 | 1·3228 | 350 | 771·618 |
| 0·7 | 1·5432 | 400 | 881·849 |
| 0·8 | 1·7637 | 450 | 992·080 |
| 0·9 | 1·9842 | 500 | 1 102·311 |
| | | 550 | 1 212·542 |
| 1 | 2·2046 | 600 | 1 322·774 |
| 2 | 4·4092 | 650 | 1 433·005 |
| 3 | 6·6139 | 700 | 1 543·236 |
| 4 | 8·8185 | 750 | 1 653·467 |
| 5 | 11·0231 | 800 | 1 763·698 |
| 6 | 13·2277 | 850 | 1 873·929 |
| 7 | 15·4324 | 900 | 1 984·160 |
| 8 | 17·6370 | 950 | 2 094·391 |
| 9 | 19·8416 | | |
| | | 1 000 | 2 204·62 |
| 10 | 22·0462 | 1 500 | 3 306·93 |
| 15 | 33·0693 | 2 000 | 4 409·25 |
| 20 | 44·0925 | 2 500 | 5 511·56 |
| 25 | 55·1156 | 3 000 | 6 613·87 |
| 30 | 66·1387 | 3 500 | 7 716·18 |
| 35 | 77·1618 | 4 000 | 8 818·49 |
| 40 | 88·1849 | 4 500 | 9 920·80 |
| 45 | 99·2080 | 5 000 | 11 023·11 |
| 50 | 110·2311 | 5 500 | 12 125·42 |
| 55 | 121·2542 | 6 000 | 13 227·74 |
| 60 | 132·2774 | 6 500 | 14 330·05 |
| 65 | 143·3005 | 7 000 | 15 432·36 |
| 70 | 154·3236 | 7 500 | 16 534·67 |
| 75 | 165·3467 | 8 000 | 17 636·98 |
| 80 | 176·3698 | 8 500 | 18 739·29 |
| 85 | 187·3929 | 9 000 | 19 841·60 |
| 90 | 198·4160 | 9 500 | 20 943·91 |
| 95 | 209·4391 | 10 000 | 22 046·23 |

| LENGTH | yards (yd) to metres (m) | | |
|---|---|---|---|
| yd | m | yd | m |
| $\frac{1}{8}$ | 0·1143 | 130 | 118·872 |
| $\frac{1}{4}$ | 0·2286 | 140 | 128·016 |
| $\frac{3}{8}$ | 0·3429 | 150 | 137·160 |
| $\frac{1}{2}$ | 0·4572 | 160 | 146·304 |
| $\frac{5}{8}$ | 0·5715 | 170 | 155·448 |
| $\frac{3}{4}$ | 0·6858 | 180 | 164·592 |
| $\frac{7}{8}$ | 0·8001 | 190 | 173·736 |
| 1 | 0·9144 | 200 | 182·880 |
| 2 | 1·8288 | 210 | 192·024 |
| 3 | 2·7432 | 220 | 201·168 |
| 4 | 3·6576 | 230 | 210·312 |
| 5 | 4·5720 | 240 | 219·456 |
| 6 | 5·4864 | 250 | 228·600 |
| 7 | 6·4008 | 260 | 237·744 |
| 8 | 7·3152 | 270 | 246·888 |
| 9 | 8·2296 | 280 | 256·032 |
| | | 290 | 265·176 |
| 10 | 9·1440 | | |
| 15 | 13·7160 | 300 | 274·320 |
| 20 | 18·2880 | 310 | 283·464 |
| 25 | 22·8600 | 320 | 292·608 |
| 30 | 27·4320 | 330 | 301·752 |
| 35 | 32·0040 | 340 | 310·896 |
| 40 | 36·5760 | 350 | 320·040 |
| 45 | 41·1480 | 360 | 329·184 |
| 50 | 45·7200 | 370 | 338·328 |
| 55 | 50·2920 | 380 | 347·472 |
| 60 | 54·8640 | 390 | 356·616 |
| 65 | 59·4360 | | |
| 70 | 64·0080 | 400 | 365·760 |
| 75 | 68·5800 | 410 | 374·904 |
| 80 | 73·1520 | 420 | 384·048 |
| 85 | 77·7240 | 430 | 393·192 |
| 90 | 82·2960 | 440 | 402·336 |
| 95 | 86·8680 | 450 | 411·480 |
| | | 460 | 420·624 |
| 100 | 91·4400 | 470 | 429·768 |
| 110 | 100·5840 | 480 | 438·912 |
| 120 | 109·7280 | 490 | 448·056 |

| yards (yd) to metres (m) | | LENGTH | |
| --- | --- | --- | --- |
| yd | m | yd | m |
| 500 | 457·200 | 850 | 777·240 |
| 510 | 466·344 | 860 | 786·384 |
| 520 | 475·488 | 870 | 795·528 |
| 530 | 484·632 | 880 | 804·672 |
| 540 | 493·776 | 890 | 813·816 |
| 550 | 502·920 | | |
| 560 | 512·064 | 900 | 822·960 |
| 570 | 521·208 | 910 | 832·104 |
| 580 | 530·352 | 920 | 841·248 |
| 590 | 539·496 | 930 | 850·392 |
| | | 940 | 859·536 |
| 600 | 548·640 | 950 | 868·680 |
| 610 | 557·784 | 960 | 877·824 |
| 620 | 566·928 | 970 | 886·968 |
| 630 | 576·072 | 980 | 896·112 |
| 640 | 585·216 | 990 | 905·256 |
| 650 | 594·360 | | |
| 660 | 603·504 | 1 000 | 914·40 |
| 670 | 612·648 | 1 500 | 1 371·60 |
| 680 | 621·792 | 2 000 | 1 828·80 |
| 690 | 630·936 | 2 500 | 2 286·00 |
| | | 3 000 | 2 743·20 |
| 700 | 640·080 | 3 500 | 3 200·40 |
| 710 | 649·224 | 4 000 | 3 657·60 |
| 720 | 658·368 | 4 500 | 4 114·80 |
| 730 | 667·512 | 5 000 | 4 572·00 |
| 740 | 676·656 | 5 500 | 5 029·20 |
| 750 | 685·800 | 6 000 | 5 486·40 |
| 760 | 694·944 | 6 500 | 5 943·60 |
| 770 | 704·088 | 7 000 | 6 400·80 |
| 780 | 713·232 | 7 500 | 6 858·00 |
| 790 | 722·376 | 8 000 | 7 315·20 |
| | | 8 500 | 7 772·40 |
| 800 | 731·520 | 9 000 | 8 229·60 |
| 810 | 740·664 | 9 500 | 8 686·80 |
| 820 | 749·808 | | |
| 830 | 758·952 | 10 000 | 9 144·00 |
| 840 | 768·096 | | |

| LENGTH | miles to kilometres (km) | | |
|---|---|---|---|
| miles | km | miles | km |
| $\frac{1}{8}$ | 0·2012 | 130 | 209·215 |
| $\frac{1}{4}$ | 0·4023 | 140 | 225·308 |
| $\frac{3}{8}$ | 0·6035 | 150 | 241·402 |
| $\frac{1}{2}$ | 0·8047 | 160 | 257·495 |
| $\frac{5}{8}$ | 1·0058 | 170 | 273·588 |
| $\frac{3}{4}$ | 1·2070 | 180 | 289·682 |
| $\frac{7}{8}$ | 1·4082 | 190 | 305·775 |
| | | | |
| 1 | 1·6093 | 200 | 321·869 |
| 2 | 3·2187 | 210 | 337·962 |
| 3 | 4·8280 | 220 | 354·056 |
| 4 | 6·4374 | 230 | 370·149 |
| 5 | 8·0467 | 240 | 386·243 |
| 6 | 9·6561 | 250 | 402·336 |
| 7 | 11·2654 | 260 | 418·429 |
| 8 | 12·8748 | 270 | 434·523 |
| 9 | 14·4841 | 280 | 450·616 |
| | | 290 | 466·710 |
| 10 | 16·0934 | | |
| 15 | 24·1402 | 300 | 482·803 |
| 20 | 32·1869 | 310 | 498·897 |
| 25 | 40·2336 | 320 | 514·990 |
| 30 | 48·2803 | 330 | 531·084 |
| 35 | 56·3270 | 340 | 547·177 |
| 40 | 64·3738 | 350 | 563·270 |
| 45 | 72·4205 | 360 | 579·364 |
| 50 | 80·4672 | 370 | 595·457 |
| 55 | 88·5139 | 380 | 611·551 |
| 60 | 96·5606 | 390 | 627·644 |
| 65 | 104·6074 | | |
| 70 | 112·6541 | 400 | 643·738 |
| 75 | 120·7008 | 410 | 659·831 |
| 80 | 128·7475 | 420 | 675·924 |
| 85 | 136·7942 | 430 | 692·018 |
| 90 | 144·8410 | 440 | 708·111 |
| 95 | 152·8877 | 450 | 724·205 |
| | | 460 | 740·298 |
| 100 | 160·9344 | 470 | 756·392 |
| 110 | 177·0278 | 480 | 772·485 |
| 120 | 193·1213 | 490 | 788·579 |

**Note.** This table may be used to convert miles per hour to kilometres per hour.

| miles to kilometres (km) | | | LENGTH |
|---|---|---|---|
| miles | km | miles | km |
| 500 | 804·672 | 850 | 1 367·942 |
| 510 | 820·765 | 860 | 1 384·036 |
| 520 | 836·859 | 870 | 1 400·129 |
| 530 | 852·952 | 880 | 1 416·223 |
| 540 | 869·046 | 890 | 1 432·316 |
| 550 | 885·139 | | |
| 560 | 901·233 | 900 | 1 448·410 |
| 570 | 917·326 | 910 | 1 464·503 |
| 580 | 933·420 | 920 | 1 480·596 |
| 590 | 949·513 | 930 | 1 496·690 |
| | | 940 | 1 512·783 |
| 600 | 965·606 | 950 | 1 528·877 |
| 610 | 981·700 | 960 | 1 544·970 |
| 620 | 997·793 | 970 | 1 561·064 |
| 630 | 1 013·887 | 980 | 1 577·157 |
| 640 | 1 029·980 | 990 | 1 593·251 |
| 650 | 1 046·074 | | |
| 660 | 1 062·167 | 1 000 | 1 609·34 |
| 670 | 1 078·260 | 1 500 | 2 414·02 |
| 680 | 1 094·354 | 2 000 | 3 218·69 |
| 690 | 1 110·447 | 2 500 | 4 023·36 |
| | | 3 000 | 4 828·03 |
| 700 | 1 126·541 | 3 500 | 5 632·70 |
| 710 | 1 142·634 | 4 000 | 6 437·38 |
| 720 | 1 158·728 | 4 500 | 7 242·05 |
| 730 | 1 174·821 | 5 000 | 8 046·72 |
| 740 | 1 190·915 | 5 500 | 8 851·39 |
| 750 | 1 207·008 | 6 000 | 9 656·06 |
| 760 | 1 223·101 | 6 500 | 10 460·74 |
| 770 | 1 239·195 | 7 000 | 11 265·41 |
| 780 | 1 255·288 | 7 500 | 12 070·08 |
| 790 | 1 271·382 | 8 000 | 12 874·75 |
| | | 8 500 | 13 679·42 |
| 800 | 1 287·475 | 9 000 | 14 484·10 |
| 810 | 1 303·569 | 9 500 | 15 288·77 |
| 820 | 1 319·662 | | |
| 830 | 1 335·756 | 10 000 | 16 093·44 |
| 840 | 1 351·849 | | |

| LENGTH | metres (m) to yards (yd), feet (ft), inches (in) | | | | | | |
|--------|------|-----|---------|------|------|-----|---------|
| m | yd | ft | in | m | yd | ft | in |
| 0·1 | 0 | 0 | 3·937 | 120 | 131 | 0 | 8·409 |
| 0·2 | 0 | 0 | 7·874 | 130 | 142 | 0 | 6·110 |
| 0·3 | 0 | 0 | 11·811 | 140 | 153 | 0 | 3·811 |
| 0·4 | 0 | 1 | 3·748 | 150 | 164 | 0 | 1·512 |
| 0·5 | 0 | 1 | 7·685 | 160 | 174 | 2 | 11·213 |
| 0·6 | 0 | 1 | 11·622 | 170 | 185 | 2 | 8·913 |
| 0·7 | 0 | 2 | 3·559 | 180 | 196 | 2 | 6·614 |
| 0·8 | 0 | 2 | 7·496 | 190 | 207 | 2 | 4·315 |
| 0·9 | 0 | 2 | 11·433 | | | | |
| | | | | 200 | 218 | 2 | 2·016 |
| 1 | 1 | 0 | 3·370 | 210 | 229 | 1 | 11·717 |
| 2 | 2 | 0 | 6·740 | 220 | 240 | 1 | 9·417 |
| 3 | 3 | 0 | 10·110 | 230 | 251 | 1 | 7·118 |
| 4 | 4 | 1 | 1·480 | 240 | 262 | 1 | 4·819 |
| 5 | 5 | 1 | 4·850 | 250 | 273 | 1 | 2·520 |
| 6 | 6 | 1 | 8·220 | 260 | 284 | 1 | 0·220 |
| 7 | 7 | 1 | 11·591 | 270 | 295 | 0 | 9·921 |
| 8 | 8 | 2 | 2·961 | 280 | 306 | 0 | 7·622 |
| 9 | 9 | 2 | 6·331 | 290 | 317 | 0 | 5·323 |
| 10 | 10 | 2 | 9·701 | 300 | 328 | 0 | 3·024 |
| 15 | 16 | 1 | 2·551 | 310 | 339 | 0 | 0·724 |
| 20 | 21 | 2 | 7·402 | 320 | 349 | 2 | 10·425 |
| 25 | 27 | 1 | 0·252 | 330 | 360 | 2 | 8·126 |
| 30 | 32 | 2 | 5·102 | 340 | 371 | 2 | 5·827 |
| 35 | 38 | 0 | 9·953 | 350 | 382 | 2 | 3·528 |
| 40 | 43 | 2 | 2·803 | 360 | 393 | 2 | 1·228 |
| 45 | 49 | 0 | 7·654 | 370 | 404 | 1 | 10·929 |
| 50 | 54 | 2 | 0·504 | 380 | 415 | 1 | 8·630 |
| 55 | 60 | 0 | 5·354 | 390 | 426 | 1 | 6·331 |
| 60 | 65 | 1 | 10·205 | | | | |
| 65 | 71 | 0 | 3·055 | 400 | 437 | 1 | 4·031 |
| 70 | 76 | 1 | 7·906 | 410 | 448 | 1 | 1·732 |
| 75 | 82 | 0 | 0·756 | 420 | 459 | 0 | 11·433 |
| 80 | 87 | 1 | 5·606 | 430 | 470 | 0 | 9·134 |
| 85 | 92 | 2 | 10·457 | 440 | 481 | 0 | 6·835 |
| 90 | 98 | 1 | 3·307 | 450 | 492 | 0 | 4·535 |
| 95 | 103 | 2 | 8·157 | 460 | 503 | 0 | 2·236 |
| | | | | 470 | 513 | 2 | 11·937 |
| 100 | 109 | 1 | 1·008 | 480 | 524 | 2 | 9·638 |
| 110 | 120 | 0 | 10·709 | 490 | 535 | 2 | 7·339 |

| metres (m) to yards (yd), feet (ft), inches (in) | | | | | LENGTH | | |
|---|---|---|---|---|---|---|---|
| m | yd | ft | in | m | yd | ft | in |
| 500 | 546 | 2 | 5·039 | 850 | 929 | 1 | 8·567 |
| 510 | 557 | 2 | 2·740 | 860 | 940 | 1 | 6·268 |
| 520 | 568 | 2 | 0·441 | 870 | 951 | 1 | 3·969 |
| 530 | 579 | 1 | 10·142 | 880 | 962 | 1 | 1·669 |
| 540 | 590 | 1 | 7·843 | 890 | 973 | 0 | 11·370 |
| 550 | 601 | 1 | 5·543 | | | | |
| 560 | 612 | 1 | 3·244 | 900 | 984 | 0 | 9·071 |
| 570 | 623 | 1 | 0·945 | 910 | 995 | 0 | 6·772 |
| 580 | 634 | 0 | 10·646 | 920 | 1 006 | 0 | 4·472 |
| 590 | 645 | 0 | 8·346 | 930 | 1 017 | 0 | 2·173 |
| | | | | 940 | 1 027 | 2 | 11·874 |
| 600 | 656 | 0 | 6·047 | 950 | 1 038 | 2 | 9·575 |
| 610 | 667 | 0 | 3·748 | 960 | 1 049 | 2 | 7·276 |
| 620 | 678 | 0 | 1·449 | 970 | 1 060 | 2 | 4·976 |
| 630 | 688 | 2 | 11·150 | 980 | 1 071 | 2 | 2·677 |
| 640 | 699 | 2 | 8·850 | 990 | 1 082 | 2 | 0·378 |
| 650 | 710 | 2 | 6·551 | | | | |
| 660 | 721 | 2 | 4·252 | 1 000 | 1 093 | 1 | 10·079 |
| 670 | 732 | 2 | 1·953 | 1 500 | 1 640 | 1 | 3·118 |
| 680 | 743 | 1 | 11·654 | 2 000 | 2 187 | 0 | 8·157 |
| 690 | 754 | 1 | 9·354 | 2 500 | 2 734 | 0 | 1·197 |
| | | | | 3 000 | 3 280 | 2 | 6·236 |
| 700 | 765 | 1 | 7·055 | 3 500 | 3 827 | 1 | 11·276 |
| 710 | 776 | 1 | 4·756 | 4 000 | 4 374 | 1 | 4·315 |
| 720 | 787 | 1 | 2·457 | 4 500 | 4 921 | 0 | 9·354 |
| 730 | 798 | 1 | 0·157 | 5 000 | 5 468 | 0 | 2·394 |
| 740 | 809 | 0 | 9·858 | 5 500 | 6 014 | 2 | 7·433 |
| 750 | 820 | 0 | 7·559 | 6 000 | 6 561 | 2 | 0·472 |
| 760 | 831 | 0 | 5·260 | 6 500 | 7 108 | 1 | 5·512 |
| 770 | 842 | 0 | 2·961 | 7 000 | 7 655 | 1 | 10·551 |
| 780 | 853 | 0 | 0·661 | 7 500 | 8 202 | 0 | 3·591 |
| 790 | 863 | 2 | 10·362 | 8 000 | 8 748 | 2 | 8·630 |
| | | | | 8 500 | 9 295 | 2 | 1·669 |
| 800 | 874 | 2 | 8·063 | 9 000 | 9 842 | 1 | 6·709 |
| 810 | 885 | 2 | 5·764 | 9 500 | 10 389 | 0 | 11·748 |
| 820 | 896 | 2 | 3·465 | | | | |
| 830 | 907 | 2 | 1·165 | 10 000 | 10 936 | 0 | 4·787 |
| 840 | 918 | 1 | 10·866 | | | | |

# LENGTH — kilometres (km) to miles, yards (yd)

| km | miles | yd | km | miles | yd |
|---|---|---|---|---|---|
| 0·1 | 0 | 109·36 | 120 | 74 | 993·60 |
| 0·2 | 0 | 218·72 | 130 | 80 | 1 369·73 |
| 0·3 | 0 | 328·08 | 140 | 86 | 1 745·86 |
| 0·4 | 0 | 437·45 | 150 | 93 | 361·99 |
| 0·5 | 0 | 546·81 | 160 | 99 | 738·13 |
| 0·6 | 0 | 656·17 | 170 | 105 | 1 114·26 |
| 0·7 | 0 | 765·53 | 180 | 111 | 1 490·39 |
| 0·8 | 0 | 874·89 | 190 | 118 | 106·53 |
| 0·9 | 0 | 984·25 | | | |
| | | | 200 | 124 | 482·66 |
| 1 | 0 | 1 093·61 | 210 | 130 | 858·79 |
| 2 | 1 | 427·23 | 220 | 136 | 1 234·93 |
| 3 | 1 | 1 520·84 | 230 | 142 | 1 611·06 |
| 4 | 2 | 854·45 | 240 | 149 | 227·19 |
| 5 | 3 | 188·07 | 250 | 155 | 603·32 |
| 6 | 3 | 1 281·68 | 260 | 161 | 979·46 |
| 7 | 4 | 615·29 | 270 | 167 | 1 355·59 |
| 8 | 4 | 1 708·91 | 280 | 173 | 1 731·72 |
| 9 | 5 | 1 042·52 | 290 | 180 | 347·86 |
| | | | | | |
| 10 | 6 | 376·13 | 300 | 186 | 723·99 |
| 15 | 9 | 564·20 | 310 | 192 | 1 100·12 |
| 20 | 12 | 752·27 | 320 | 198 | 1 476·26 |
| 25 | 15 | 940·33 | 330 | 205 | 92·39 |
| 30 | 18 | 1 128·40 | 340 | 211 | 468·52 |
| 35 | 21 | 1 316·47 | 350 | 217 | 844·65 |
| 40 | 24 | 1 504·53 | 360 | 223 | 1 220·79 |
| 45 | 27 | 1 692·60 | 370 | 229 | 1 596·92 |
| 50 | 31 | 120·66 | 380 | 236 | 213·05 |
| 55 | 34 | 308·73 | 390 | 242 | 589·19 |
| 60 | 37 | 496·80 | | | |
| 65 | 40 | 684·86 | 400 | 248 | 965·32 |
| 70 | 43 | 872·93 | 410 | 254 | 1 341·45 |
| 75 | 46 | 1 061·00 | 420 | 260 | 1 717·59 |
| 80 | 49 | 1 249·06 | 430 | 267 | 333·72 |
| 85 | 52 | 1 437·13 | 440 | 273 | 709·85 |
| 90 | 55 | 1 625·20 | 450 | 279 | 1 085·98 |
| 95 | 59 | 853·26 | 460 | 285 | 1 462·12 |
| | | | 470 | 292 | 78·25 |
| 100 | 62 | 241·33 | 480 | 298 | 454·38 |
| 110 | 68 | 617·46 | 490 | 304 | 830·52 |

| km | miles | yd | km | miles | yd |
|---|---|---|---|---|---|
| 500 | 310 | 1 206·65 | 850 | 528 | 291·30 |
| 510 | 316 | 1 582·78 | 860 | 534 | 667·44 |
| 520 | 323 | 198·92 | 870 | 540 | 1 043·57 |
| 530 | 329 | 575·05 | 880 | 546 | 1 419·70 |
| 540 | 335 | 951·18 | 890 | 553 | 35·84 |
| 550 | 341 | 1 327·31 | | | |
| 560 | 347 | 1 703·45 | 900 | 559 | 411·97 |
| 570 | 354 | 319·58 | 910 | 565 | 788·10 |
| 580 | 360 | 695·71 | 920 | 571 | 1 164·23 |
| 590 | 366 | 1 071·85 | 930 | 577 | 1 540·37 |
| | | | 940 | 584 | 156·50 |
| 600 | 372 | 1 447·98 | 950 | 590 | 532·63 |
| 610 | 379 | 64·11 | 960 | 596 | 908·77 |
| 620 | 385 | 440·24 | 970 | 602 | 1 284·90 |
| 630 | 391 | 816·38 | 980 | 608 | 1 661·03 |
| 640 | 397 | 1 192·51 | 990 | 615 | 277·17 |
| 650 | 403 | 1 568·64 | | | |
| 660 | 410 | 184·78 | 1 000 | 621 | 653·30 |
| 670 | 416 | 560·91 | 1 500 | 932 | 99·95 |
| 680 | 422 | 937·04 | 2 000 | 1 242 | 1 306·60 |
| 690 | 428 | 1 313·18 | 2 500 | 1 553 | 753·25 |
| | | | 3 000 | 1 864 | 199·90 |
| 700 | 434 | 1 689·31 | 3 500 | 2 174 | 1 406·54 |
| 710 | 441 | 305·44 | 4 000 | 2 485 | 853·19 |
| 720 | 447 | 681·57 | 4 500 | 2 796 | 299·84 |
| 730 | 453 | 1 057·71 | 5 000 | 3 106 | 1 506·49 |
| 740 | 459 | 1 433·84 | 5 500 | 3 417 | 953·14 |
| 750 | 466 | 49·97 | 6 000 | 3 728 | 399·79 |
| 760 | 472 | 426·11 | 6 500 | 4 038 | 1 606·44 |
| 770 | 478 | 802·24 | 7 000 | 4 349 | 1 053·09 |
| 780 | 484 | 1 178·37 | 7 500 | 4 660 | 499·74 |
| 790 | 490 | 1 554·51 | 8 000 | 4 970 | 1 706·39 |
| | | | 8 500 | 5 281 | 1 153·04 |
| 800 | 497 | 170·64 | 9 000 | 5 592 | 599·69 |
| 810 | 503 | 546·77 | 9 500 | 5 903 | 46·33 |
| 820 | 509 | 922·90 | | | |
| 830 | 515 | 1 299·04 | 10 000 | 6 213 | 1 252·98 |
| 840 | 521 | 1 675·17 | | | |

| CAPACITY | | gallons (gal) to litres (l) | |
|---|---|---|---|
| gal | l | gal | l |
| 1/16 | 0·5682 | 130 | 590·976 |
| 1/8 | 1·1365 | 140 | 636·436 |
| 1/4 | 1·7047 | 150 | 681·896 |
| 3/8 | 2·2730 | 160 | 727·356 |
| 1/2 | 2·8412 | 170 | 772·815 |
| 5/8 | 3·4095 | 180 | 818·275 |
| 3/4 | 3·9777 | 190 | 863·735 |
| 7/8 | | | |
| 1 | 4·5460 | 200 | 909·194 |
| 2 | 9·0919 | 210 | 954·654 |
| 3 | 13·6379 | 220 | 1 000·114 |
| 4 | 18·1839 | 230 | 1 045·574 |
| 5 | 22·7299 | 240 | 1 091·033 |
| 6 | 27·2758 | 250 | 1 136·493 |
| 7 | 31·8218 | 260 | 1 181·953 |
| 8 | 36·3678 | 270 | 1 227·412 |
| 9 | 40·9137 | 280 | 1 272·872 |
| | | 290 | 1 318·332 |
| 10 | 45·4597 | | |
| 15 | 68·1896 | 300 | 1 363·792 |
| 20 | 90·9194 | 310 | 1 409·251 |
| 25 | 113·6493 | 320 | 1 454·711 |
| 30 | 136·3792 | 330 | 1 500·171 |
| 35 | 159·1090 | 340 | 1 545·630 |
| 40 | 181·8389 | 350 | 1 591·090 |
| 45 | 204·5687 | 360 | 1 636·550 |
| 50 | 227·2986 | 370 | 1 682·010 |
| 55 | 250·0285 | 380 | 1 727·469 |
| 60 | 272·7583 | 390 | 1 772·929 |
| 65 | 295·4882 | | |
| 70 | 318·2180 | 400 | 1 818·389 |
| 75 | 340·9479 | 410 | 1 863·849 |
| 80 | 363·6778 | 420 | 1 909·308 |
| 85 | 386·4076 | 430 | 1 954·768 |
| 90 | 409·1375 | 440 | 2 000·228 |
| 95 | 431·8673 | 450 | 2 045·687 |
| | | 460 | 2 091·147 |
| 100 | 454·5972 | 470 | 2 136·607 |
| 110 | 500·0569 | 480 | 2 182·067 |
| 120 | 545·5166 | 490 | 2 227·526 |

| gallons (gal) to litres (l) | | CAPACITY | |
|---|---|---|---|
| gal | l | gal | l |
| 500 | 2 272·986 | 850 | 3 864·08 |
| 510 | 2 318·446 | 860 | 3 909·54 |
| 520 | 2 363·905 | 870 | 3 955·00 |
| 530 | 2 409·365 | 880 | 4 000·46 |
| 540 | 2 454·825 | 890 | 4 045·92 |
| 550 | 2 500·285 | | |
| 560 | 2 545·744 | 900 | 4 091·37 |
| 570 | 2 591·204 | 910 | 4 136·83 |
| 580 | 2 636·664 | 920 | 4 182·29 |
| 590 | 2 682·123 | 930 | 4 227·75 |
| | | 940 | 4 273·21 |
| 600 | 2 727·583 | 950 | 4 318·67 |
| 610 | 2 773·043 | 960 | 4 364·13 |
| 620 | 2 818·503 | 970 | 4 409·59 |
| 630 | 2 863·962 | 980 | 4 455·05 |
| 640 | 2 909·422 | 990 | 4 500·51 |
| 650 | 2 954·882 | | |
| 660 | 3 000·342 | 1 000 | 4 545·97 |
| 670 | 3 045·801 | 1 500 | 6 818·96 |
| 680 | 3 091·261 | 2 000 | 9 091·94 |
| 690 | 3 136·721 | 2 500 | 11 364·93 |
| | | 3 000 | 13 637·92 |
| 700 | 3 182·180 | 3 500 | 15 910·90 |
| 710 | 3 227·640 | 4 000 | 18 183·89 |
| 720 | 3 273·100 | 4 500 | 20 456·87 |
| 730 | 3 318·560 | 5 000 | 22 729·86 |
| 740 | 3 364·019 | 5 500 | 25 002·85 |
| 750 | 3 409·479 | 6 000 | 27 275·83 |
| 760 | 3 454·939 | 6 500 | 29 548·82 |
| 770 | 3 500·398 | 7 000 | 31 821·80 |
| 780 | 3 545·858 | 7 500 | 34 094·79 |
| 790 | 3 591·318 | 8 000 | 36 367·78 |
| | | 8 500 | 38 640·76 |
| 800 | 3 636·778 | 9 000 | 40 913·75 |
| 810 | 3 682·237 | 9 500 | 43 186·73 |
| 820 | 3 727·697 | | |
| 830 | 3 773·157 | 10 000 | 45 459·72 |
| 840 | 3 818·616 | | |

| CAPACITY | litres (l) to gallons (gal) and pints (pt) |
| --- | --- |

| l | gal | pt | l | gal | pt |
| --- | --- | --- | --- | --- | --- |
| 0·1 | 0 | 0·176 | 120 | 26 | 3·176 |
| 0·2 | 0 | 0·352 | 130 | 28 | 4·774 |
| 0·3 | 0 | 0·528 | 140 | 30 | 6·372 |
| 0·4 | 0 | 0·704 | 150 | 32 | 7·970 |
| 0·5 | 0 | 0·880 | 160 | 35 | 1·568 |
| 0·6 | 0 | 1·056 | 170 | 37 | 3·166 |
| 0·7 | 0 | 1·232 | 180 | 39 | 4·764 |
| 0·8 | 0 | 1·408 | 190 | 41 | 6·362 |
| 0·9 | 0 | 1·584 | | | |
| | | | 200 | 43 | 7·960 |
| 1 | 0 | 1·760 | 210 | 46 | 1·558 |
| 2 | 0 | 3·520 | 220 | 48 | 3·156 |
| 3 | 0 | 5·279 | 230 | 50 | 4·754 |
| 4 | 0 | 7·039 | 240 | 52 | 6·352 |
| 5 | 1 | 0·799 | 250 | 54 | 7·950 |
| 6 | 1 | 2·559 | 260 | 57 | 1·548 |
| 7 | 1 | 4·319 | 270 | 59 | 3·146 |
| 8 | 1 | 6·078 | 280 | 61 | 4·744 |
| 9 | 1 | 7·838 | 290 | 63 | 6·342 |
| 10 | 2 | 1·598 | 300 | 65 | 7·940 |
| 15 | 3 | 2·397 | 310 | 68 | 1·538 |
| 20 | 4 | 3·196 | 320 | 70 | 3·136 |
| 25 | 5 | 3·995 | 330 | 72 | 4·734 |
| 30 | 6 | 4·794 | 340 | 74 | 6·332 |
| 35 | 7 | 5·593 | 350 | 76 | 7·930 |
| 40 | 8 | 6·392 | 360 | 79 | 1·528 |
| 45 | 9 | 7·191 | 370 | 81 | 3·126 |
| 50 | 10 | 7·990 | 380 | 83 | 4·724 |
| 55 | 12 | 0·789 | 390 | 85 | 6·322 |
| 60 | 13 | 1·588 | | | |
| 65 | 14 | 2·387 | 400 | 87 | 7·920 |
| 70 | 15 | 3·186 | 410 | 90 | 1·518 |
| 75 | 16 | 3·985 | 420 | 92 | 3·116 |
| 80 | 17 | 4·784 | 430 | 94 | 4·714 |
| 85 | 18 | 5·583 | 440 | 96 | 6·312 |
| 90 | 19 | 6·382 | 450 | 98 | 7·910 |
| 95 | 20 | 7·181 | 460 | 101 | 1·508 |
| | | | 470 | 103 | 3·106 |
| 100 | 21 | 7·980 | 480 | 105 | 4·704 |
| 110 | 24 | 1·578 | 490 | 107 | 6·302 |

| litres (l) to gallons (gal) and pints (pt) | | | **CAPACITY** | | |
|---|---|---|---|---|---|
| l | gal | pt | l | gal | pt |
| 500 | 109 | 7·900 | 850 | 186 | 7·830 |
| 510 | 112 | 1·498 | 860 | 189 | 1·428 |
| 520 | 114 | 3·096 | 870 | 191 | 3·026 |
| 530 | 116 | 4·694 | 880 | 193 | 4·624 |
| 540 | 118 | 6·292 | 890 | 195 | 6·222 |
| 550 | 120 | 7·890 | | | |
| 560 | 123 | 1·488 | 900 | 197 | 7·820 |
| 570 | 125 | 3·086 | 910 | 200 | 1·418 |
| 580 | 127 | 4·684 | 920 | 202 | 3·016 |
| 590 | 129 | 6·282 | 930 | 204 | 4·614 |
| | | | 940 | 206 | 6·212 |
| 600 | 131 | 7·880 | 950 | 208 | 7·810 |
| 610 | 134 | 1·478 | 960 | 211 | 1·408 |
| 620 | 136 | 3·076 | 970 | 213 | 3·006 |
| 630 | 138 | 4·674 | 980 | 215 | 4·604 |
| 640 | 140 | 6·272 | 990 | 217 | 6·202 |
| 650 | 142 | 7·870 | | | |
| 660 | 145 | 1·468 | 1 000 | 219 | 7·800 |
| 670 | 147 | 3·066 | 1 500 | 329 | 7·699 |
| 680 | 149 | 4·664 | 2 000 | 439 | 7·599 |
| 690 | 151 | 6·262 | 2 500 | 549 | 7·499 |
| | | | 3 000 | 659 | 7·399 |
| 700 | 153 | 7·860 | 3 500 | 769 | 7·299 |
| 710 | 156 | 1·458 | 4 000 | 879 | 7·199 |
| 720 | 158 | 3·056 | 4 500 | 989 | 7·098 |
| 730 | 160 | 4·654 | 5 000 | 1 099 | 6·998 |
| 740 | 162 | 6·252 | 5 500 | 1 209 | 6·898 |
| 750 | 164 | 7·850 | 6 000 | 1 319 | 6·798 |
| 760 | 167 | 1·448 | 6 500 | 1 429 | 6·698 |
| 770 | 169 | 3·046 | 7 000 | 1 539 | 6·598 |
| 780 | 171 | 4·644 | 7 500 | 1 649 | 6·497 |
| 790 | 173 | 6·242 | 8 000 | 1 759 | 6·397 |
| | | | 8 500 | 1 869 | 6·297 |
| 800 | 175 | 7·840 | 9 000 | 1 979 | 6·197 |
| 810 | 178 | 1·438 | 9 500 | 2 089 | 6·097 |
| 820 | 180 | 3·036 | | | |
| 830 | 182 | 4·634 | 10 000 | 2 199 | 5·997 |
| 840 | 184 | 6·232 | | | |

| TEMPERATURE | | degrees Fahrenheit (°F) to Centigrade (°C) | |
|---|---|---|---|
| °F | °C | °F | °C |
| 0 | −17·778 | 110 | 43·333 |
| 5 | −15·000 | 115 | 46·111 |
| 10 | −12·222 | 120 | 48·889 |
| 15 | −9·444 | 125 | 51·667 |
| 20 | −6·667 | 130 | 54·444 |
| 25 | −3·889 | 135 | 57·222 |
| 30 | −1·111 | 140 | 60·000 |
| **32** | **0·000** | 145 | 62·778 |
| 35 | 1·667 | 150 | 65·556 |
| 40 | 4·444 | 155 | 68·333 |
| 45 | 7·222 | 160 | 71·111 |
| 50 | 10·000 | 165 | 73·889 |
| 55 | 12·778 | 170 | 76·667 |
| 60 | 15·556 | 175 | 79·444 |
| 65 | 18·333 | 180 | 82·222 |
| 70 | 21·111 | 185 | 85·000 |
| 75 | 23·889 | 190 | 87·778 |
| 80 | 26·667 | 195 | 90·556 |
| 85 | 29·444 | 200 | 93·333 |
| 90 | 32·222 | 205 | 96·111 |
| 95 | 35·000 | 210 | 98·889 |
| 100 | 37·778 | **212** | **100·000** |
| 105 | 40·556 | | |

| Centigrade (°C) to Fahrenheit (°F) | | | |
|---|---|---|---|
| °C | °F | °C | °F |
| **0** | **32·0** | 50 | 122·0 |
| 5 | 41·0 | 55 | 131·0 |
| 10 | 50·0 | 60 | 140·0 |
| 15 | 59·0 | 65 | 149·0 |
| 20 | 68·0 | 70 | 158·0 |
| 25 | 77·0 | 75 | 167·0 |
| 30 | 86·0 | 80 | 176·0 |
| 35 | 95·0 | 85 | 185·0 |
| 40 | 104·0 | 90 | 194·0 |
| 45 | 113·0 | 95 | 203·0 |
| | | **100** | **212·0** |